Mathematics

Numbers 1-10

Copyright © 2021 by A Devi Thangamaniam.

All right reserved. No part of this publication may be reproduced, distributed, or transmitted in any form or by any means, including photocopying, recording, or other electronic or mechanical methods, without the prior written permission of the author, except in the case of brief quotations embodied in critical reviews and certain other non-commercial uses permitted by copyright law.

Information: MiLu Children's Educational Source. www.my-willing.com

ISBN: 978 - 1 - 64999 - 870 - 5

Count numbers with pictures

Colour the pictures and count them

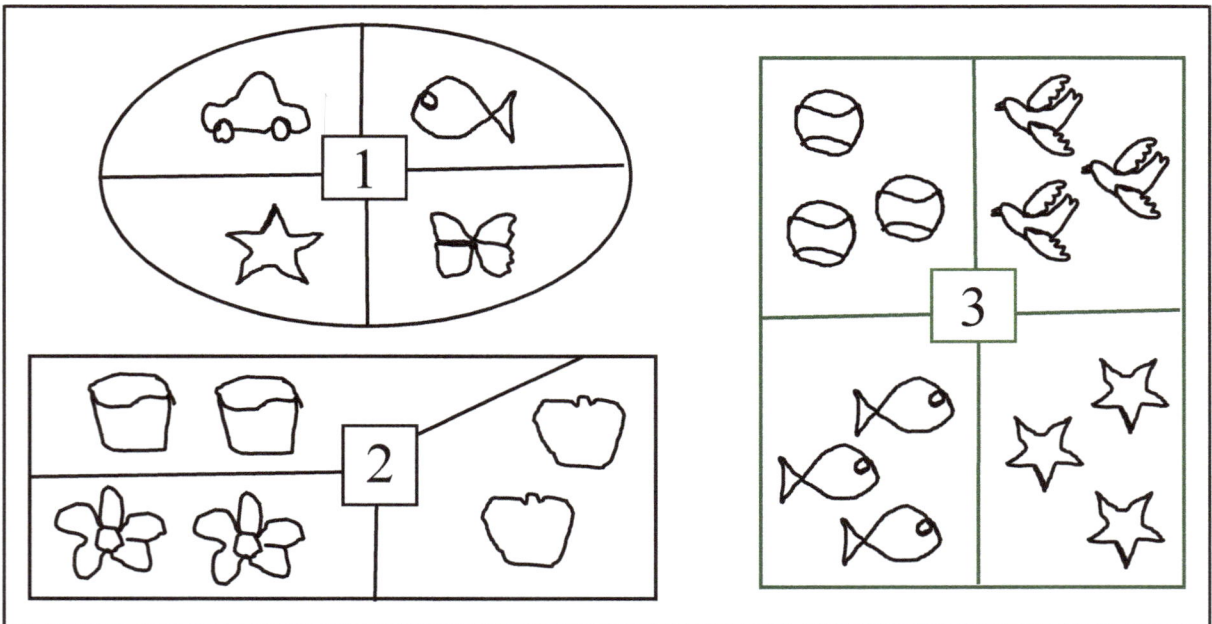

Circle the pictures for the numbers

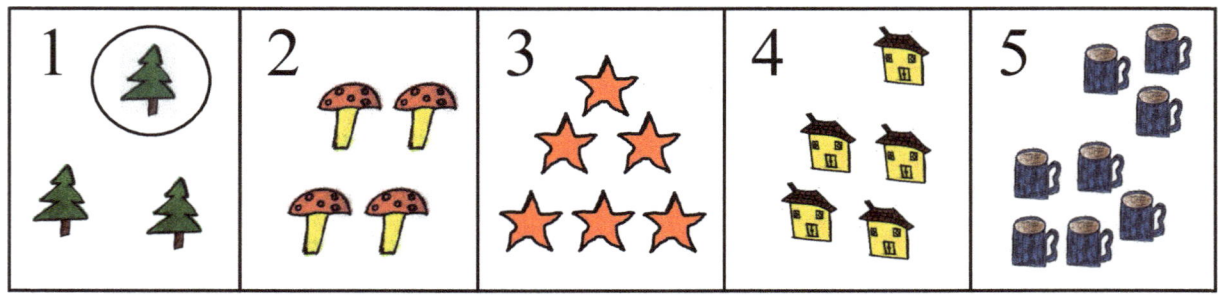

Count pictures and trace the numbers

Colour the pictures and group them

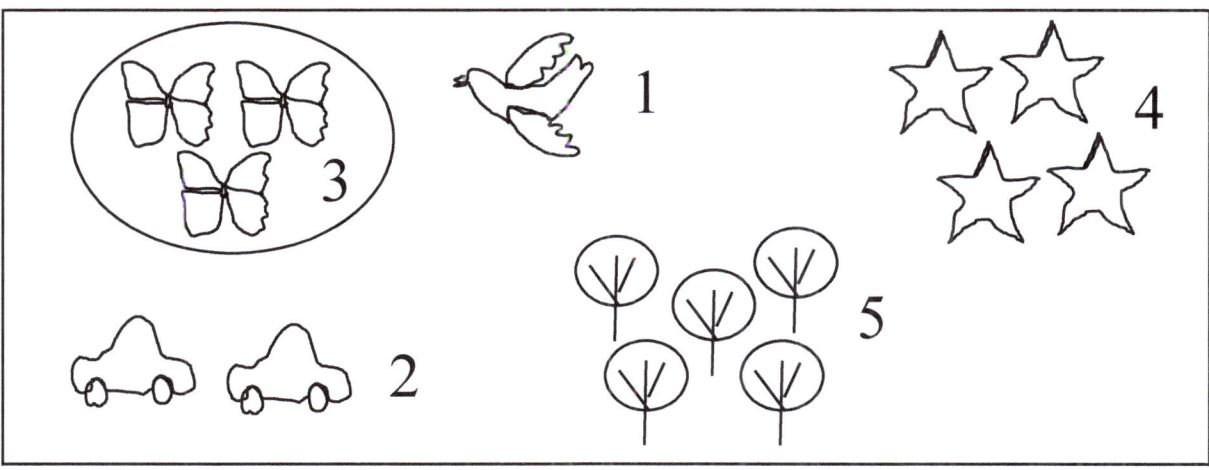

Colour the beads and circle the correct numbers

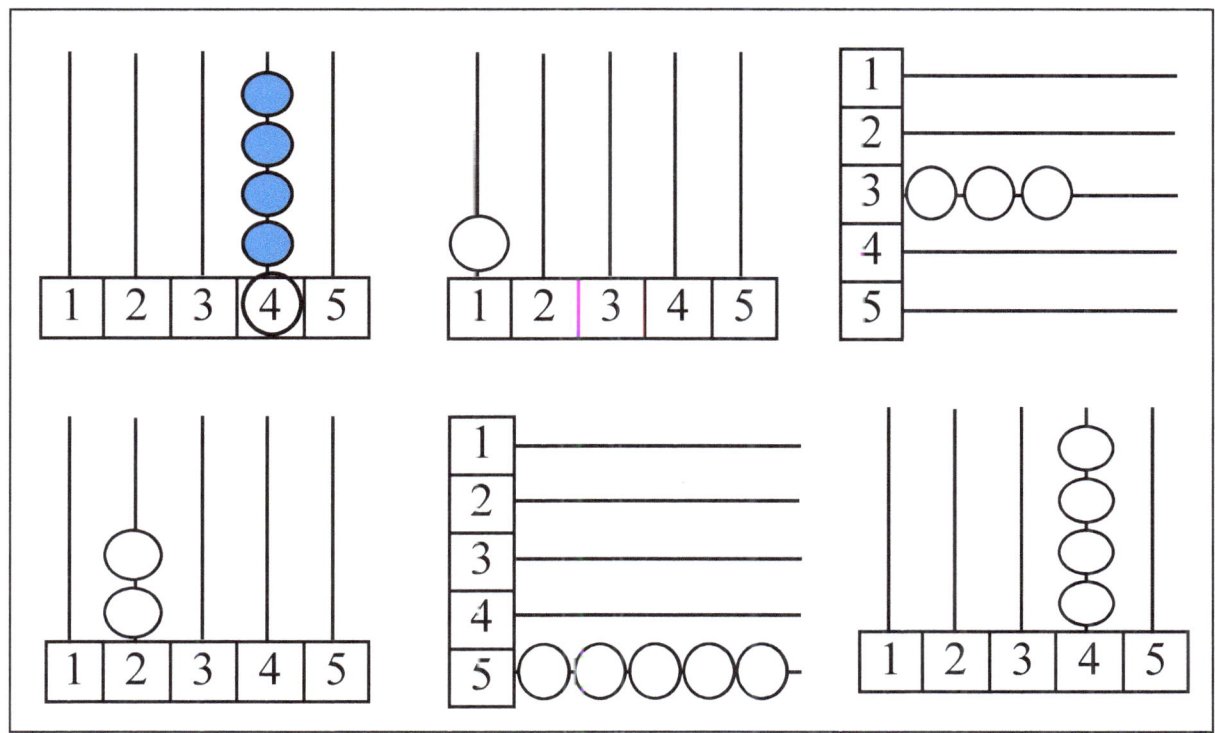

Colour the pictures and count them

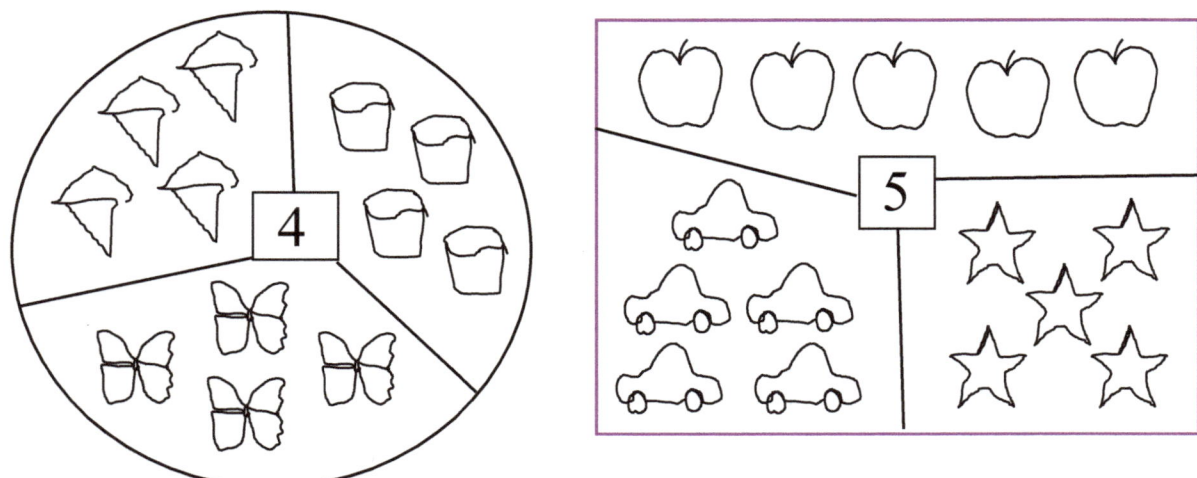

Count pictures and trace the numbers

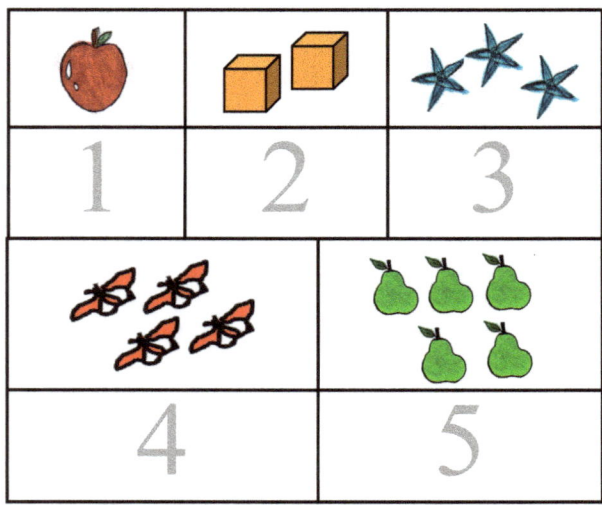

Fill in the missing numbers by pattern

Colour the pictures and count them

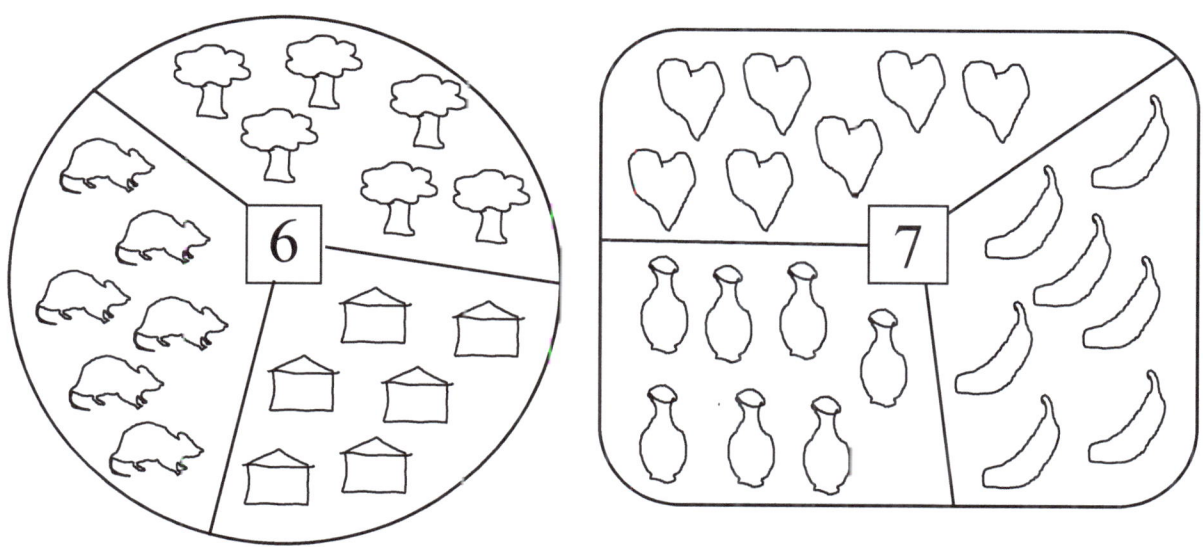

Count pictures and trace the numbers

	6	6	6	6	6	6	6
	7	7	7	7	7	7	7

Connect the numbers in order

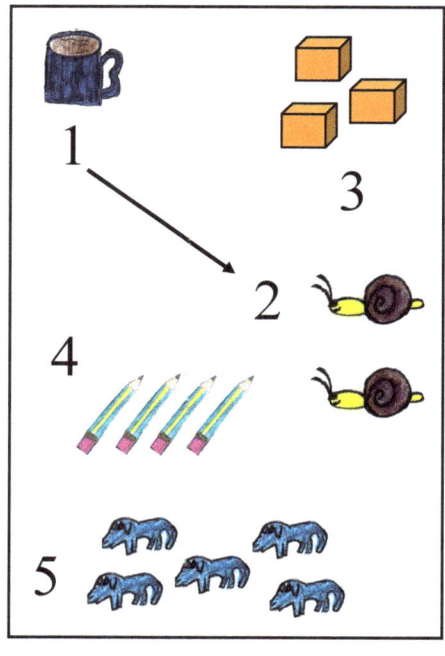
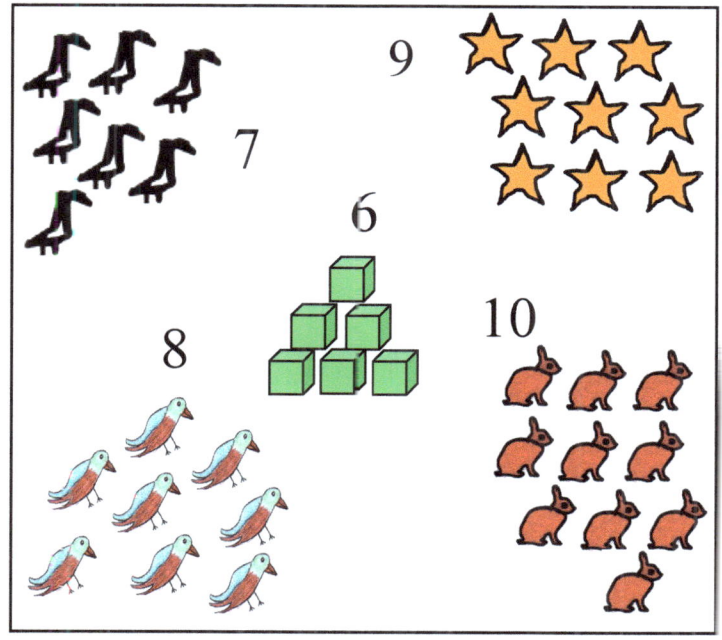

Draw pictures in the missing places and colour them

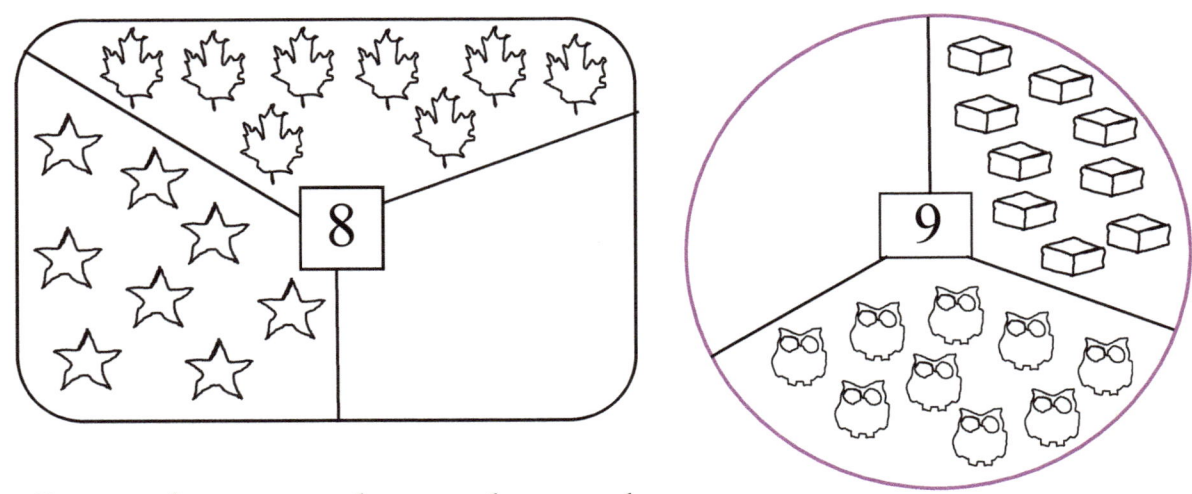

Count pictures and trace the numbers

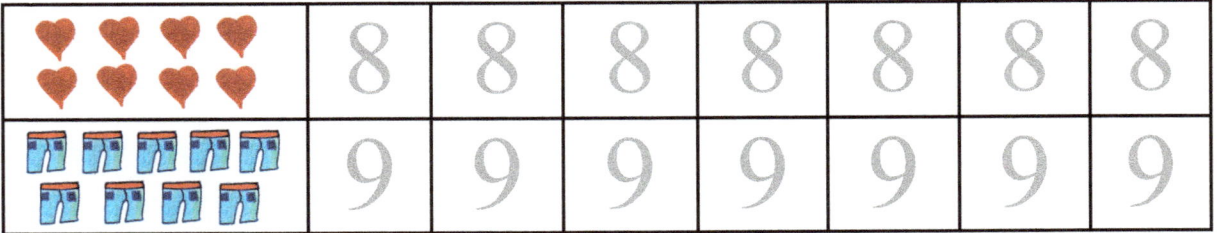

Connect the numbers in order from 1-10 and 10 - 1

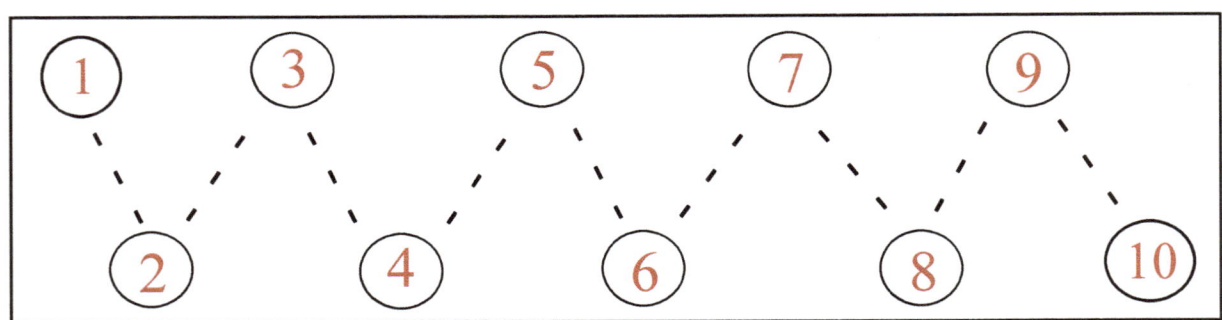

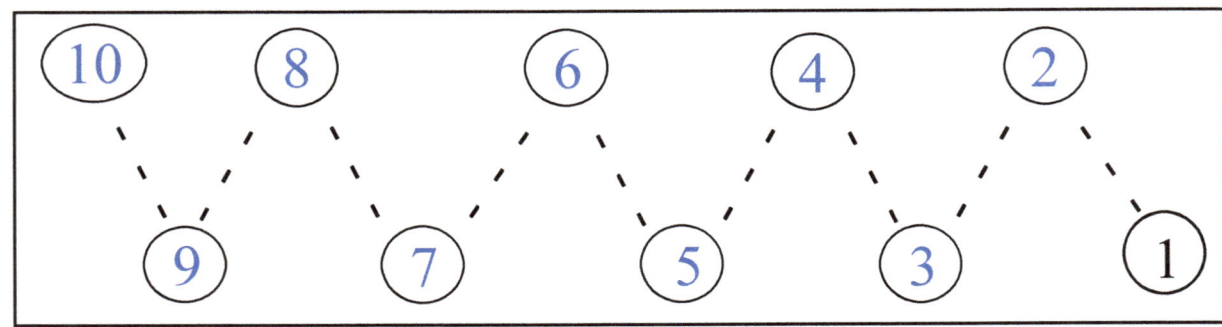

Connect these numbers with pattern

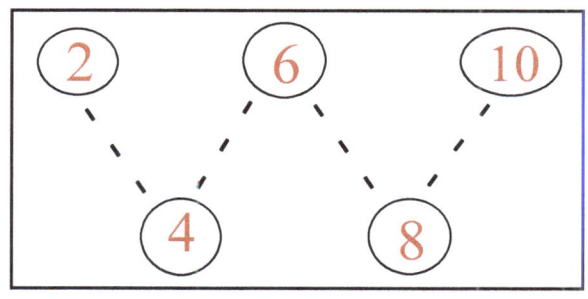
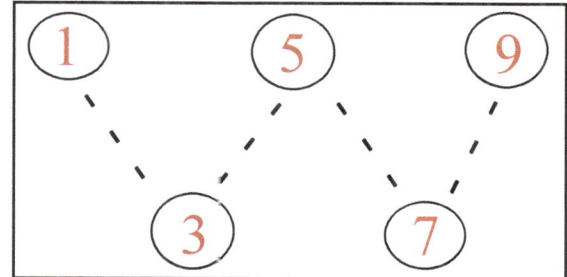
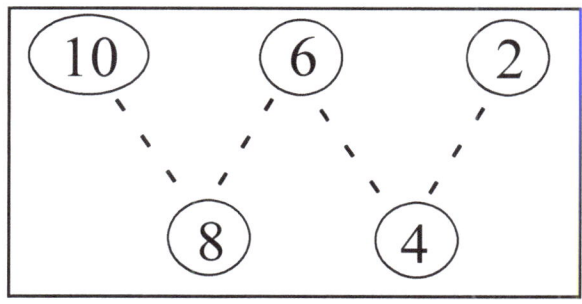
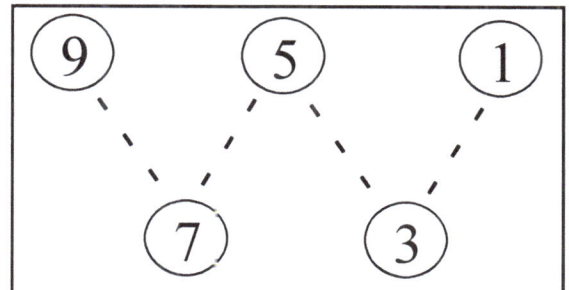

Count and match these blocks

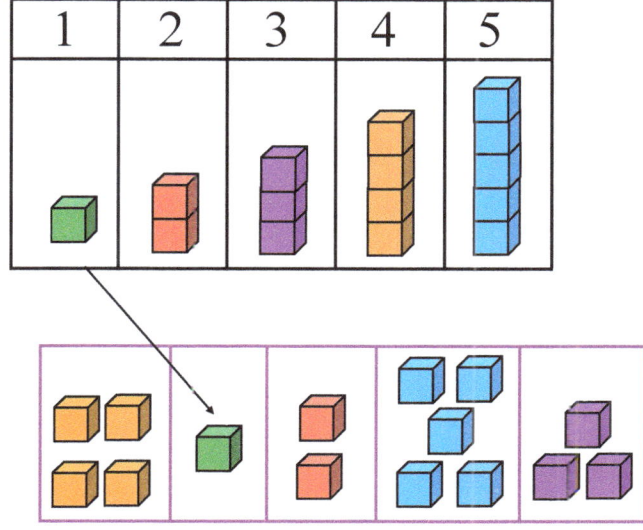
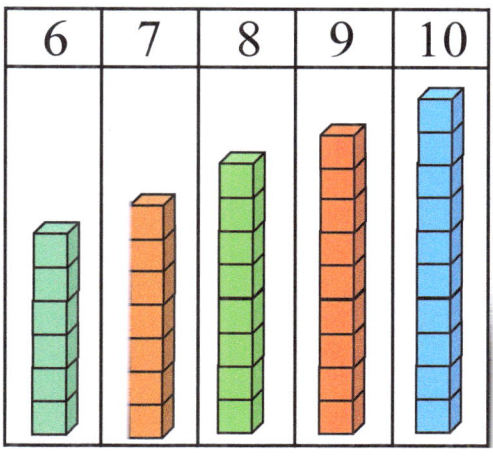
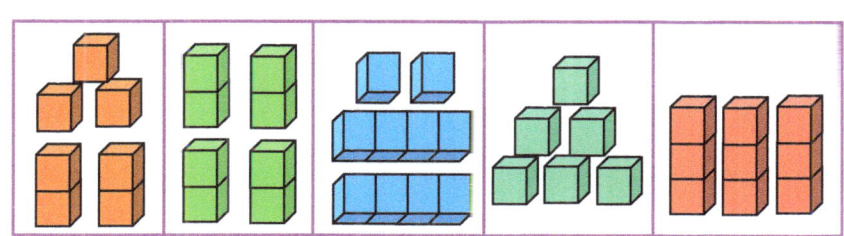

Circle the pictures for the numbers

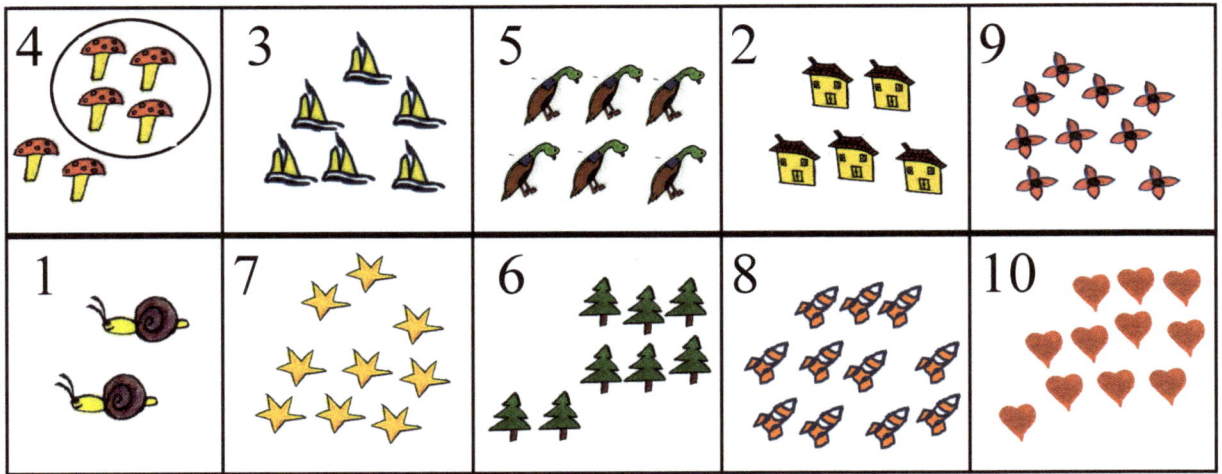

Find these numbers and circle them

1) 4, 5, 6, 7
2) 7, 8, 9, 10
3) 2, 3, 4, 5
4) 3, 4, 5, 6
5) 10, 8, 6, 4

10	9	8	7
8	2		6
6	3		5
4	4		4
6	5	4	3

Match with same numbers

7	9
10	8
9	7
6	10
8	6

1) 9, 7, 5, 3
2) 7, 8, 9, 10
3) 4, 6, 8, 10
4) 6, 5, 4, 3
5) 7, 6, 5, 4

9	7	5	3
10	8	6	4
	9		5
	10		6
7	6	5	4

2	3
3	2
4	5
1	4
5	1

Colour the pictures and trace the number

Find the same set of dices and match them

 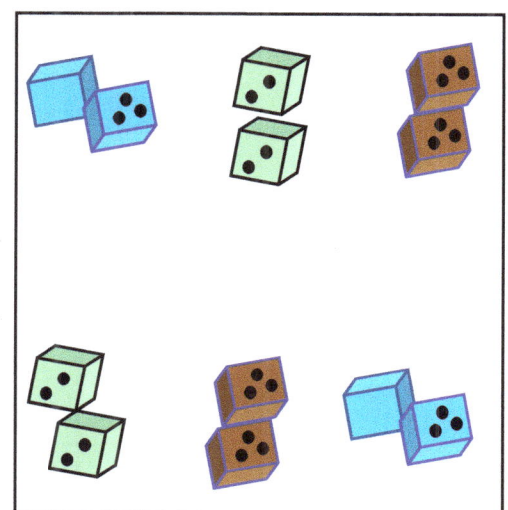

Group the same numbers

Read the numbers and colour the squares

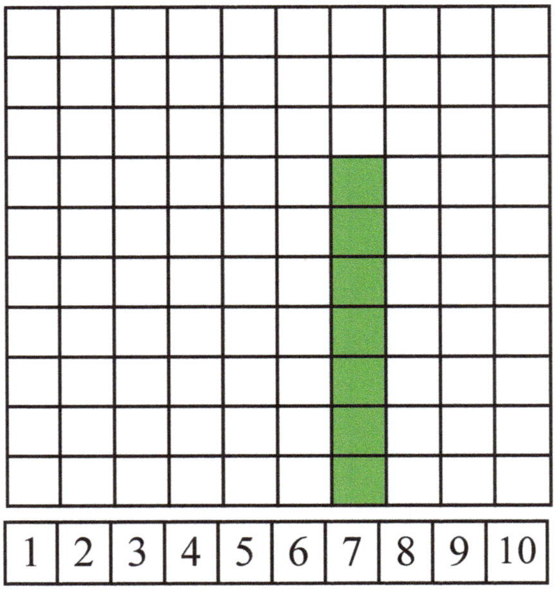

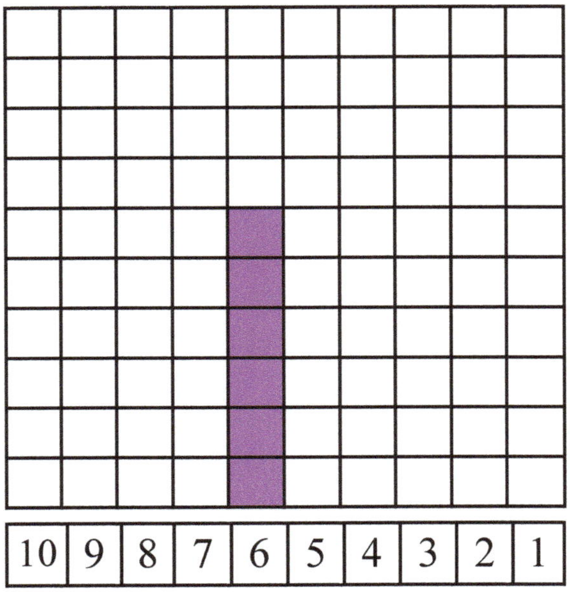

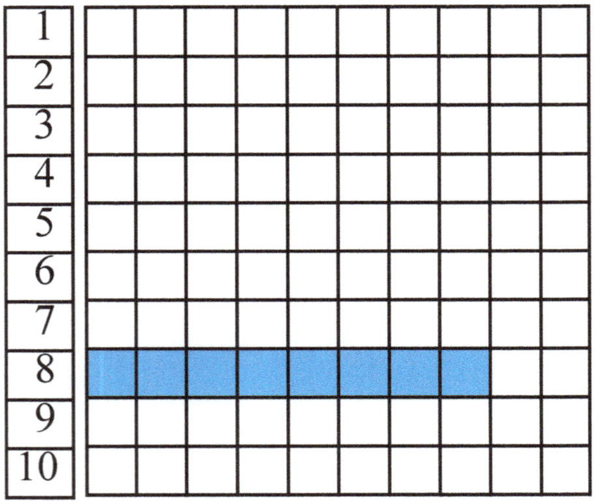

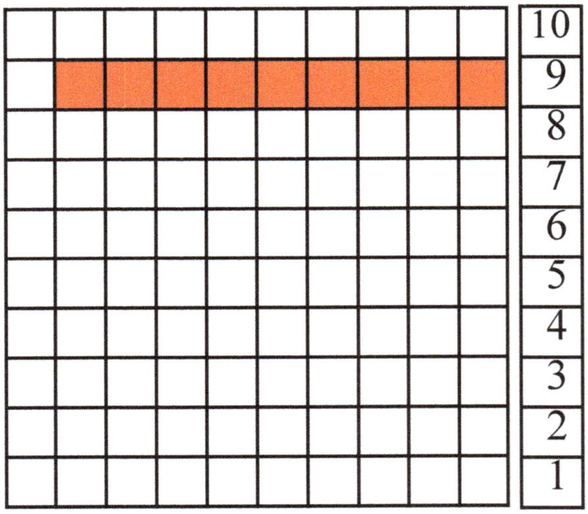

Match with same bars

Read numbers and colour the beads

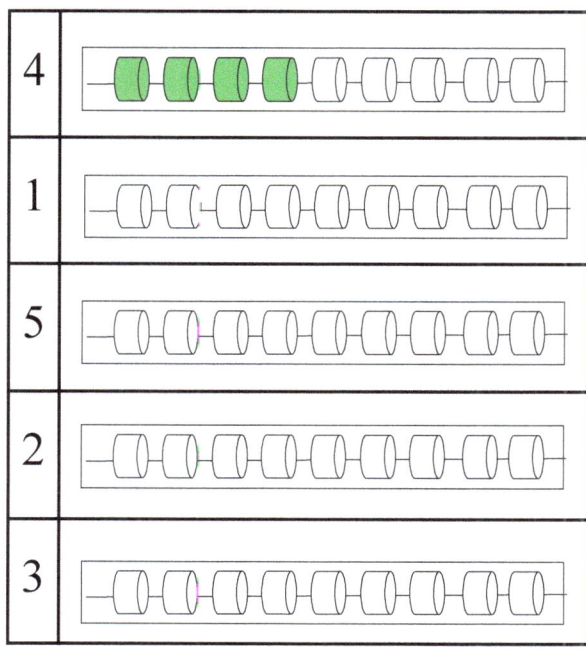

Match the previous number

2	6	4	7	8
3	7	1	5	6

Match the next number

3	9	2	5	4
6	4	10	5	3

Fill the next number in order

2, 3, 4, ___   4, 3, 2, ___   6, 7, 8, ___

4, 5, 6, ___   7, 6, 5, ___   1, 2, 3, ___

7, 8, 9, ___   9, 8, 7, ___   8, 7, 6, ___

5, 6, 7, ___   6, 5, 4, ___

3, 4, 5, ___   5, 4, 3, ___   10, 9, 8, ___

Match numbers with pictures

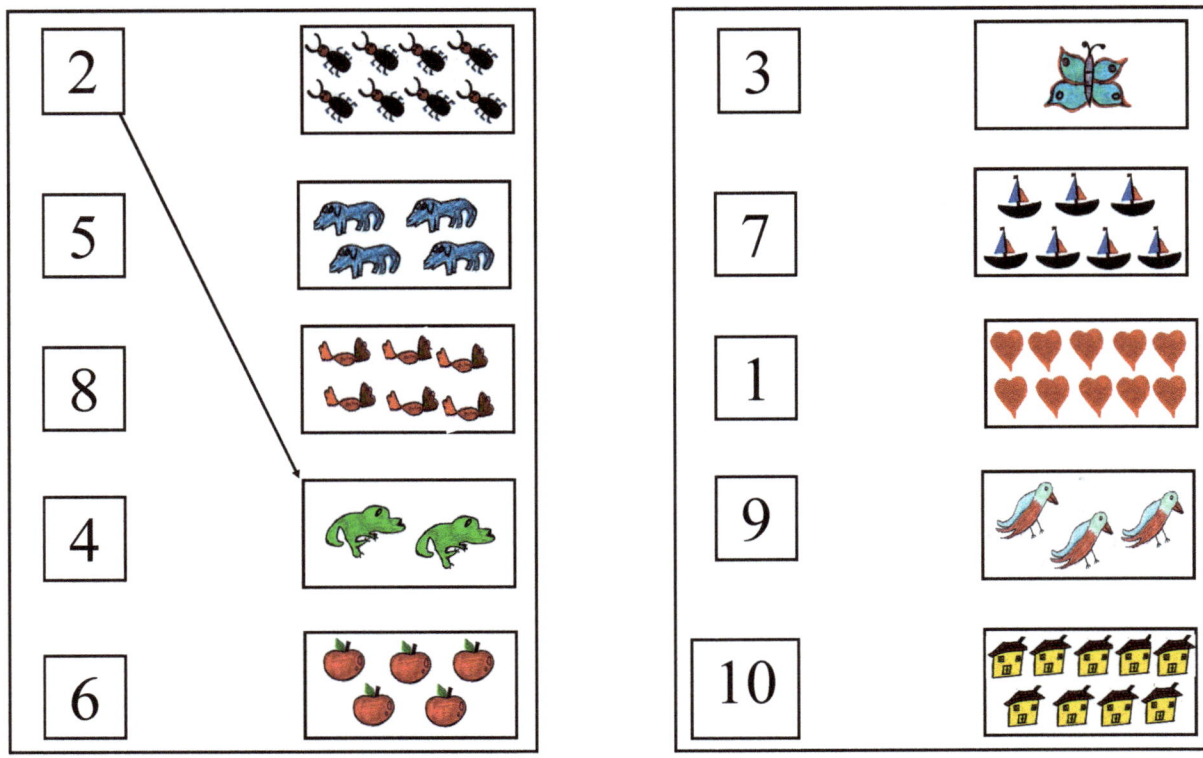

Trace the numbers 1-10 and 10-1

1	2	3	4	5	6	7	8	9	10
10	9	8	7	6	5	4	3	2	1

Colour inside the Domino and count them

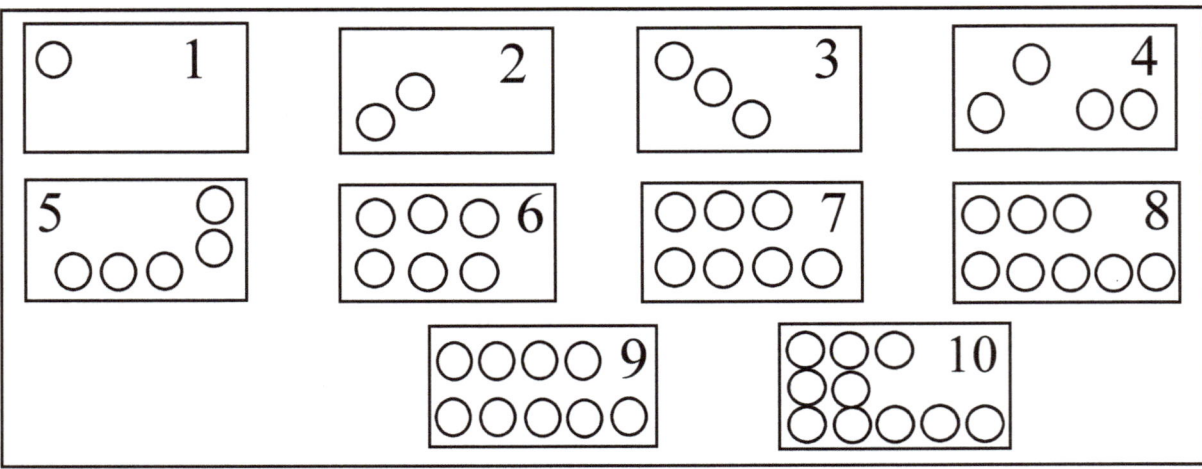

Count fingers and write the numbers in the boxes

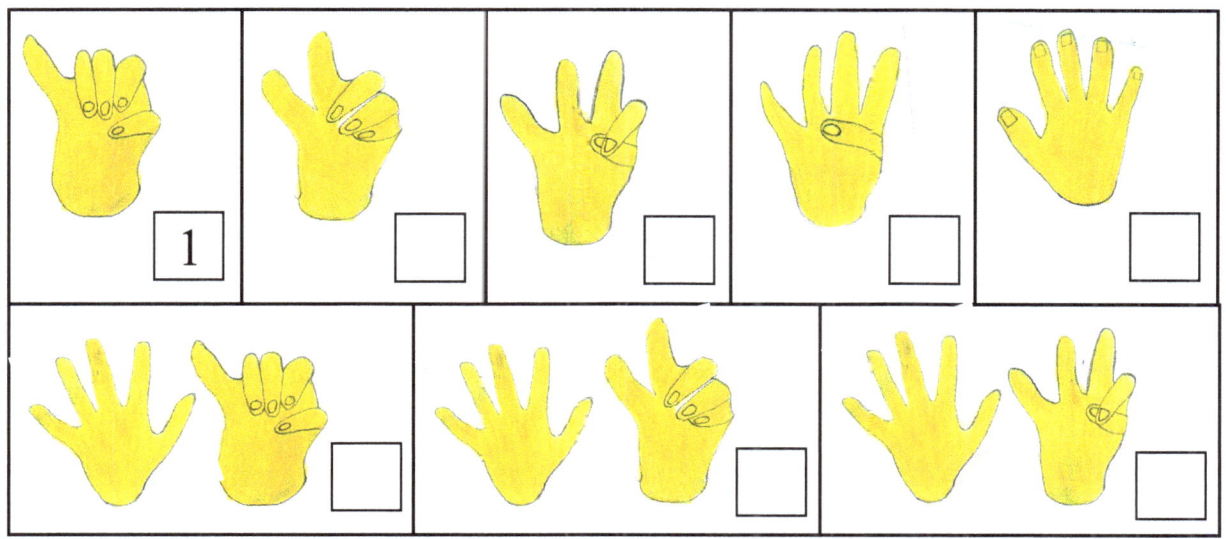

Count the toes and write the missing numbers

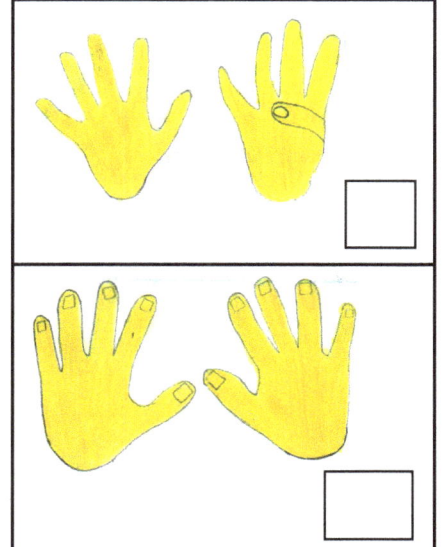

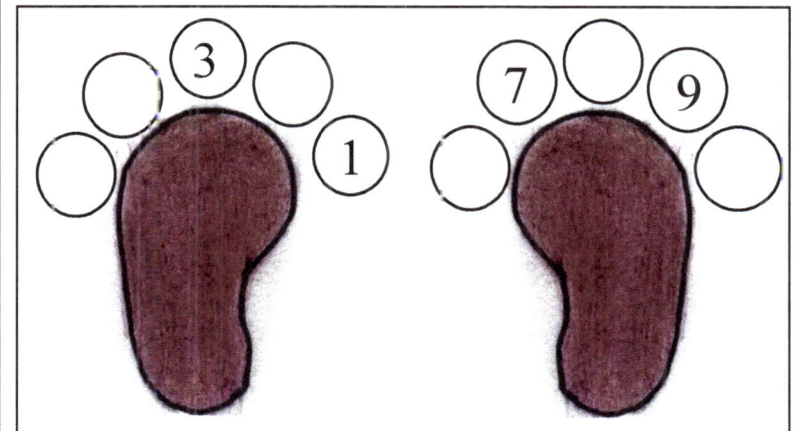

Connect the numbers in order

4 1 ↓↗↓ 5 3←2	3 5 4 6 7	10 7 8 6 9	3 2 5 4 6
7 6 9 8 5	8 7 5 6 4	5 9 1 3 7	10 4 8 6 2

Count these sticks and write the numbers in the boxes

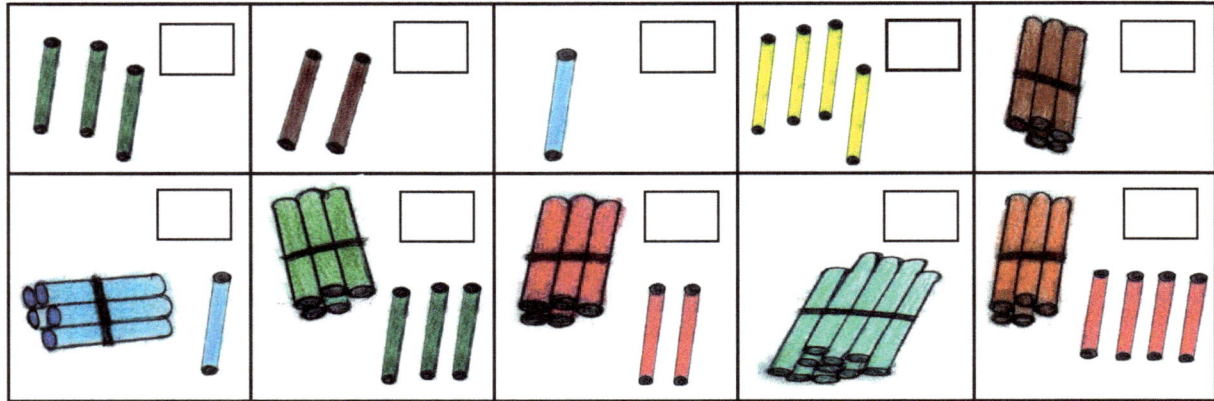

Compare and circle the smallest number

5, 3	5, 6, 4
6, 9	4, 3, 5
7, 8	7, 6, 9
8, 9	8, 10, 7
9, 6	9, 8, 6

Compare and circle the biggest number

3, 4	2, 3, 1
5, 4	3, 4, 5
9, 6	5, 4, 8
6, 9	9, 6, 8
8, 4	8, 9, 10

Count sticks and write numbers

Make the pattern with sticks from given numbers

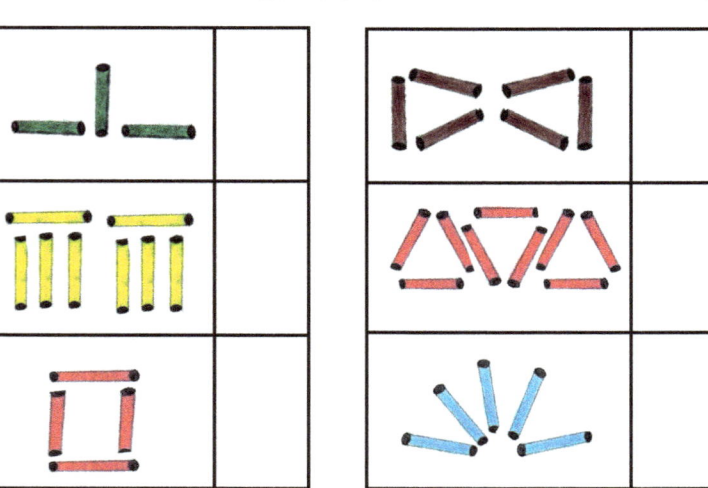

	4
	7
	5

Read the given numbers and circle them in the number line

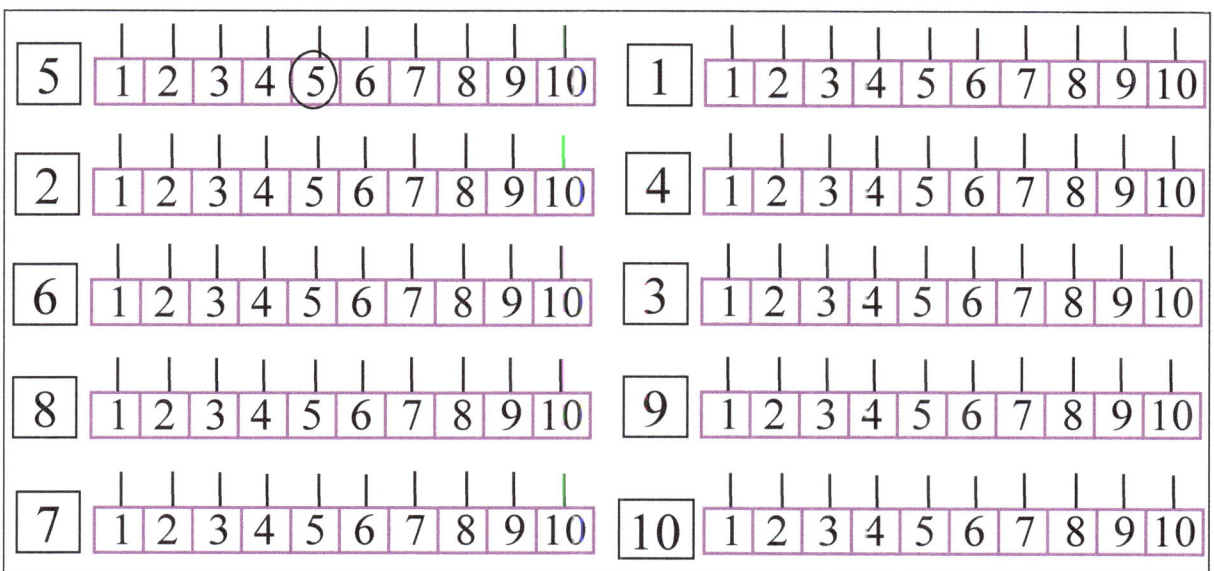

Count pictures and write numbers in the boxes

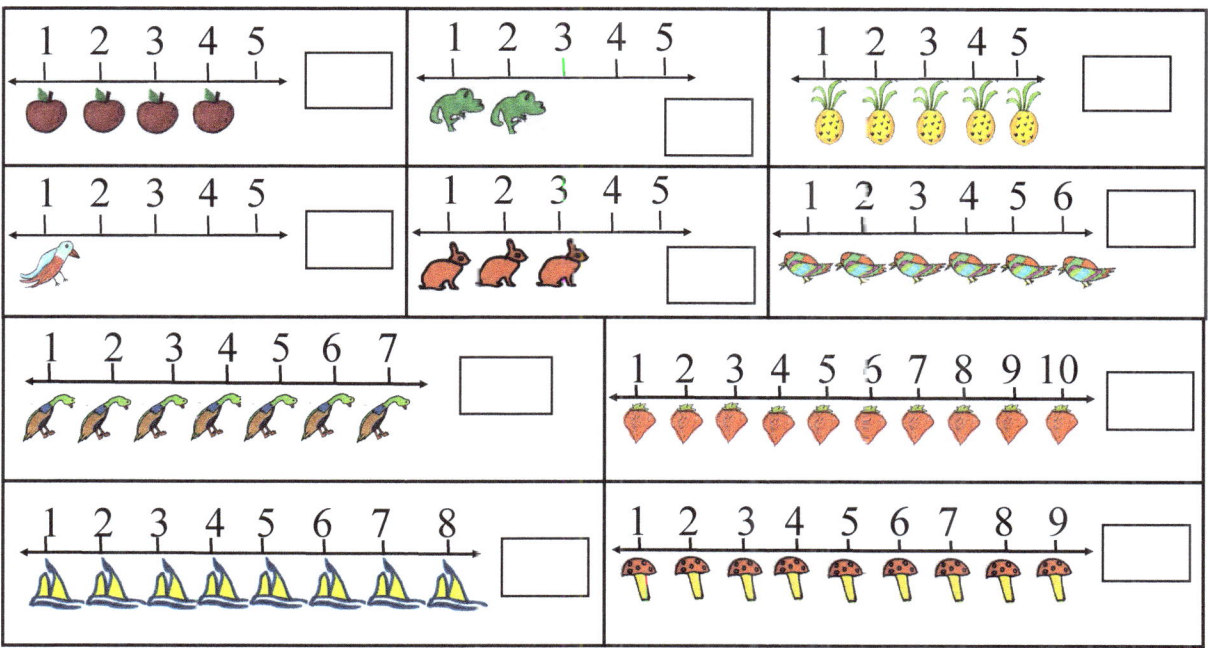

Trace the number and word

1	One	One	One	One
One	One	One	One	One

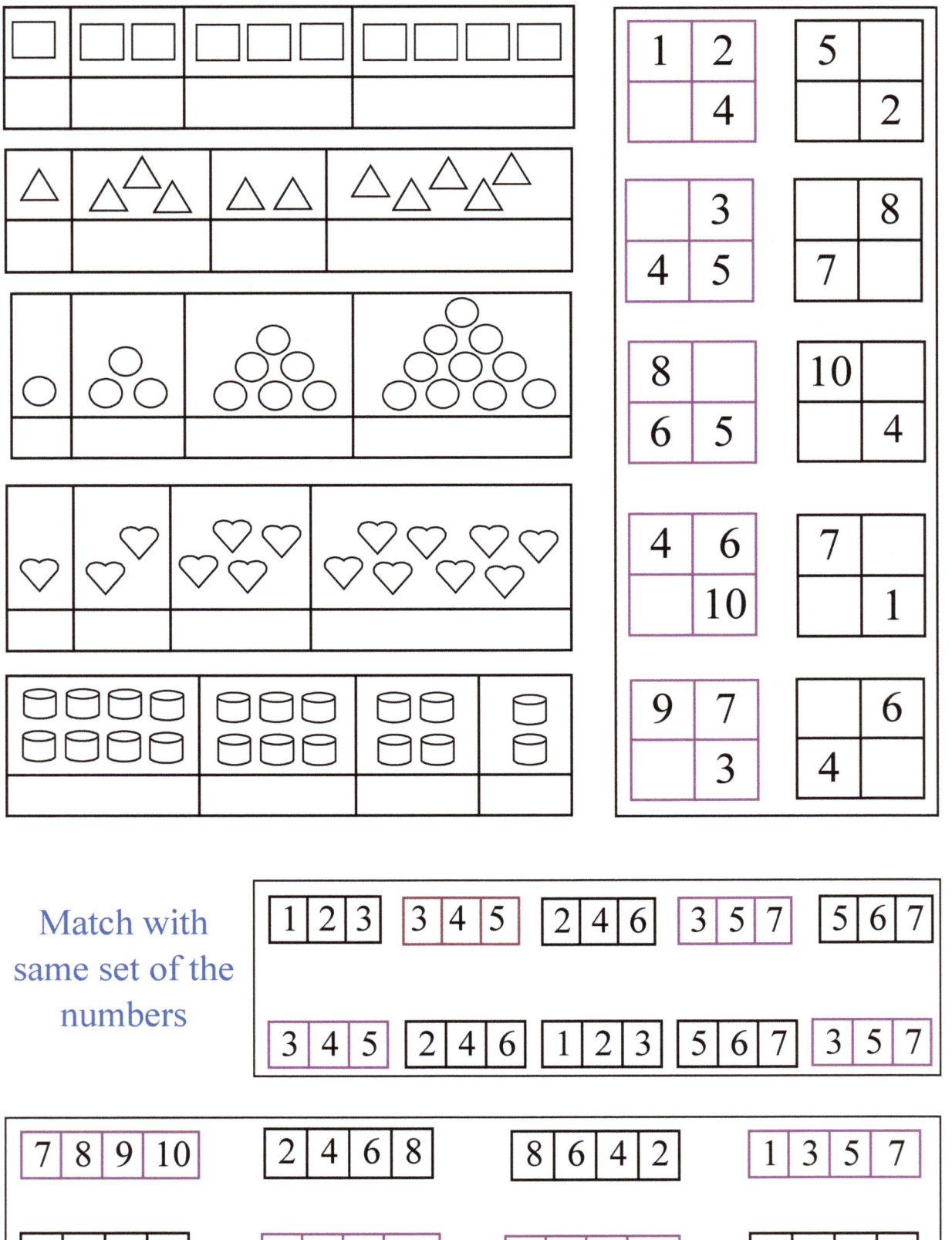

Trace the number and word

2	Two	Two	Two	Two
Two	Two	Two	Two	Two

Count pictures and circle the correct numbers

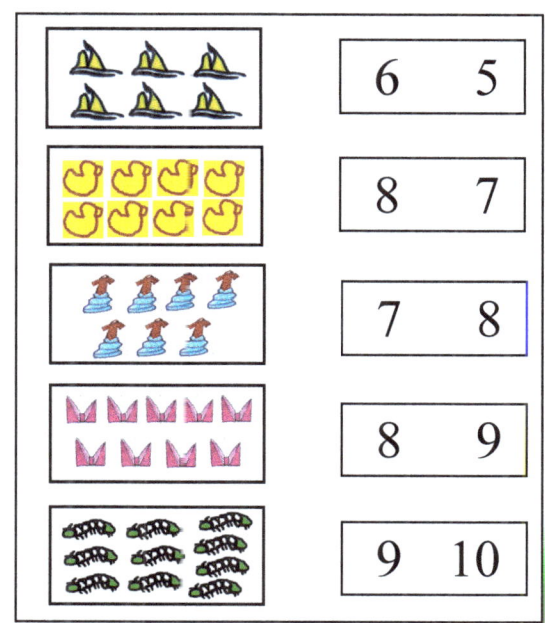

Pick up the numbers 1-10 and 10-1 and write them

2, 1, 3	1		8, 7, 6	9, 8, 10	10		8, 5, 6
2, 3, 4			4, 7, 8	8, 9, 7			4, 5, 8
1, 4, 3			9, 8, 7	8, 9, 7			4, 3, 7
4, 5, 6			6, 8, 9	6, 7, 5			6, 8, 2
6, 5, 7			9, 8, 10	7, 5, 6			3, 1, 5

Circle the same numbers

2	5	②	3
4	4	3	2
3	5	4	3

1	1	2	3
5	2	5	3
7	6	4	7

8	8	5	6
6	9	6	3
10	10	9	8
9	7	6	9

Write the before and after numbers

___, 2, ___

___, 8, ___

___, 6, ___

___, 3, ___

___, 5, ___

___, 4, ___

___, 7, ___

___, 9, ___

Circle the black coloured numbers

1	2	3	4	5
6	7	8	9	10

Circle the red coloured numbers

1	2	3	4	5
6	7	8	9	10

Fill in the missing places

1		5		9
9	7		3	

2		6	8	
10			4	2

Fill in the missing places

1,	2,	3,	4,	5
1,	__,	3,	4,	5
1,	__,	__,	4,	5
1,	__,	__,	__,	5
1,	__,	__,	__,	__
__,	__,	__,	__,	__

6,	7,	8,	9,	10
6,	__,	8,	9,	10
6,	__,	__,	9,	10
6,	__,	__,	__,	10
6,	__,	__,	__,	__
__,	__,	__,	__,	__

Climb up from 1-10 and descend from 10 - 1

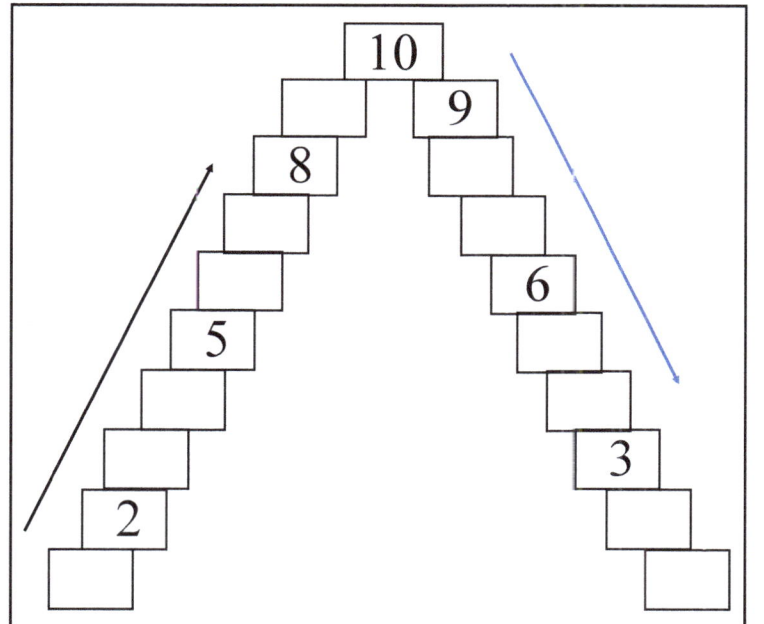
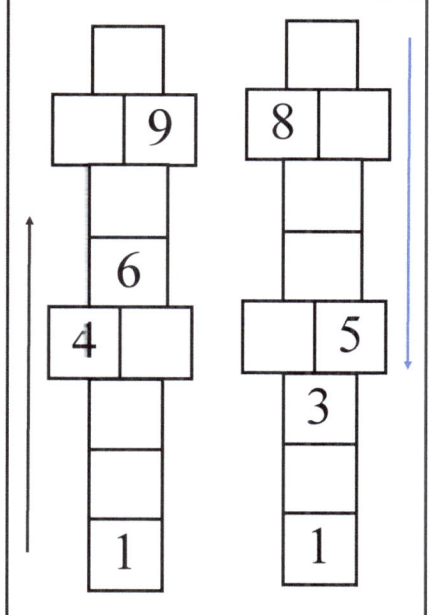

Trace the number and word

3	Three	Three	Three
Three	Three	Three	Three

Match with same words

Five	Four
One	Two
Six	Five
Two	One
Four	Six

Find the digit words and circle them

t	e	n		o	s
h		n	i	n	e
r				e	v
e			i		e
e	i	g	h	t	n
	f	i	v	e	
	o				
	u		t	w	o
	r		s	i	x

Seven	Ten
Ten	Nine
Eight	Seven
Three	Eight
Nine	Three

Fill in the missing letters and make digit words

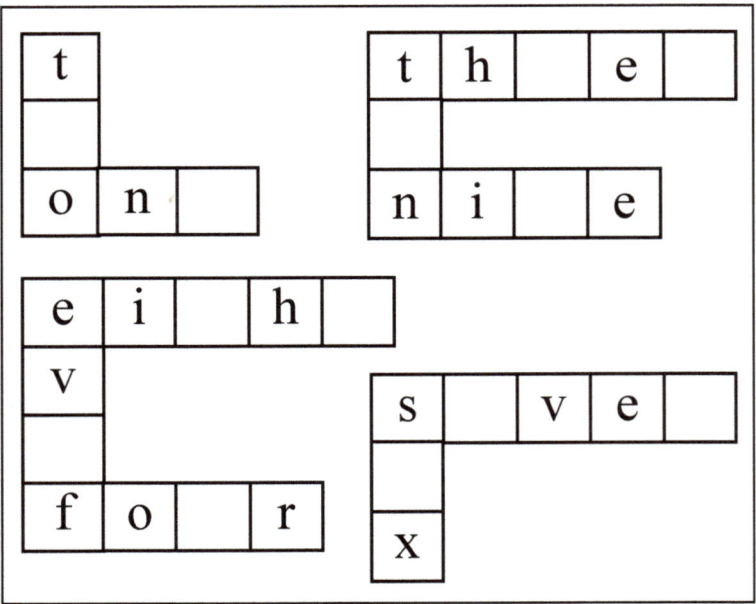

Trace the number and word

| 4 | Four | Four | Four |
| Four | Four | Four | Four |

Fill the missing letter and make digit words

1	2
O n e	T w o
O _ _ e	T _ _ o
O _ _ _	T _ _ _
_ _ _	_ _ _

3	4
T h r e e	F o u r
T _ r _ e	F _ u r
T _ r _ _	F _ _ r
T _ _ _ _	F _ _ _

Count the pictures and answer the questions

How many ? 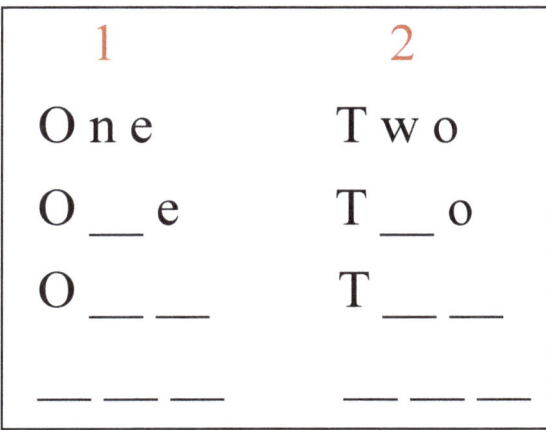 ☐
How many? ☐
How many? ☐
How many? ☐
How many? ☐
How many? ☐
How many? ☐

Arrange the numbers from big to small

1, 3, 2	3, 2, 1
5, 4, 6	__, __, __
8, 6, 7	__, __, __
9, 8, 7	__, __, __

Arrange the numbers from small to big

6, 5, 7	__, __, __
6, 4, 5	__, __, __
5, 7, 9	__, __, __
10, 6, 8	__, __, __

Fill these numbers with patterns

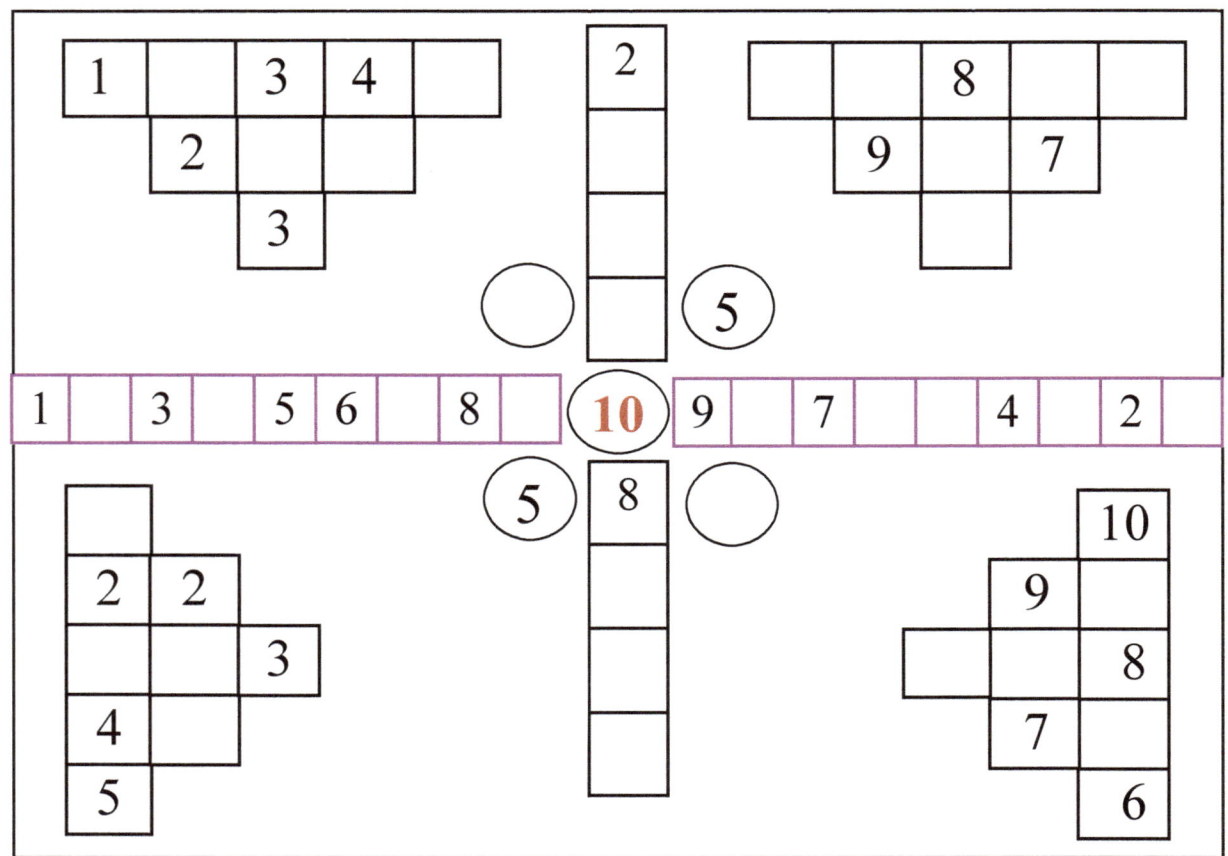

Fill in the missing letters

	5	6	7
	F i v e	S i x	S e v e n
	F _ v e	S _ x	S _ v _ n
	F _ _ e	S _ _	S _ v _ _
	F _ _ _	_ _ _	S _ _ _ _

Trace the number and words

| 5 | Five | Five | Five | Five | Five | Five |

Read words and circle the correct numbers

Four	Seven	Ten	One	Five
8 4	7 4	5 10	1 3	7 5
Three	Nine	Six	Two	Eight
4 3	9 6	6 10	2 5	5 8

Match numbers and their words with pictures

3		Four
5		Three
1		Two
4		Five
2		One

6		Eight
9		Ten
8		Nine
7		Seven
10		Six

Trace the number and word

| 6 | Six | Six | Six | Six | Six | Six |

Arrange these word in correctly

t	o	w

f	o	r	u

n	e	o

f	v	e	i

e	r	t	h	e

e	n	n	i

n	e	t

h	e	i	t	g

s	e	n	v	e

s	x	i

Find the numbers that place for pictures and circle them

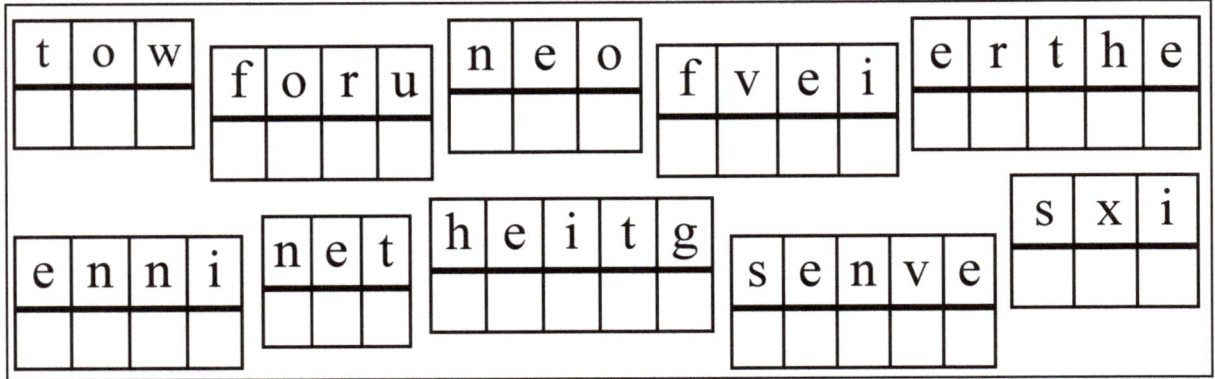

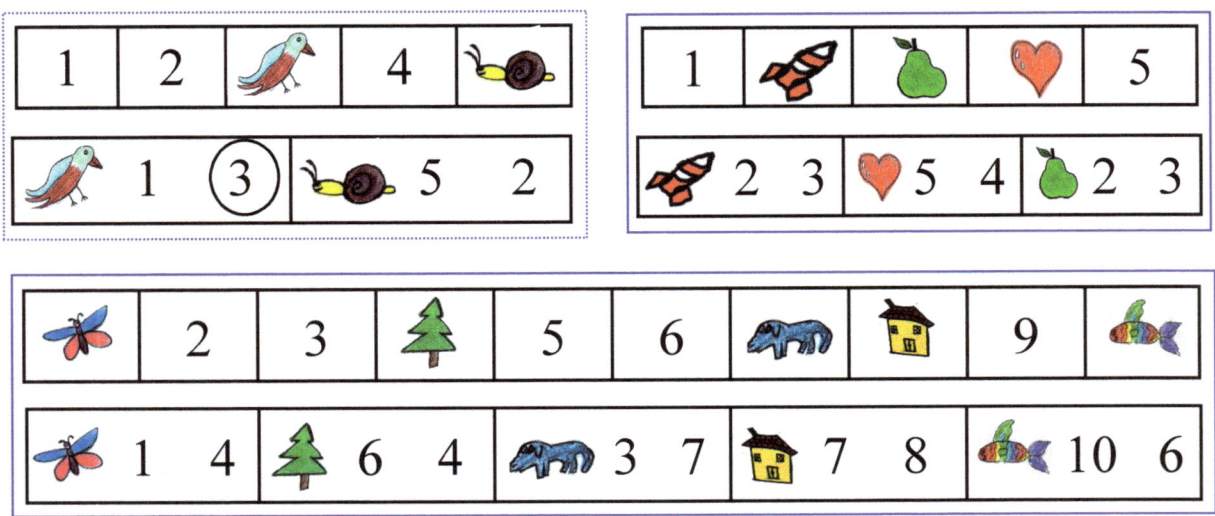

Write the numbers that place for pictures

Trace the number and word

7	Seven	Seven	Seven
Seven	Seven	Seven	Seven

Count and fill in the missing numbers and letters

🖐	|	1	O__e
🖐	||	2	Tw__
🖐	|||		Thre_
🖐	||||	4	__our
🖐	|||||		F_ve
🖐🖐	||||| |	6	__ix
🖐🖐	||||| ||		Seve_
🖐🖐	||||| |||		Ei__ht
🖐🖐	||||| ||||	9	Ni__e
🖐🖐	||||| |||||	10	T__n

Write the previous numbers

___, 4, 5
___, 2, 3
___, 7, 8
___, 8, 9

___, 3, 4
___, 5, 6
___, 9, 10
___, 6, 7

Write the middle numbers

2, ___, 4
5, ___, 7
7, ___, 9
1, ___, 3

3, ___, 5
4, ___, 6
6, ___, 8
8, ___, 10

Match them
(I have, Where are?)

I have 4 apples	I have 6 trees	I have 3 flowers	I have 7 cups
Where are 8 apples?	Where are 9 trees?	Where are 5 flowers?	Where are 10 cups?

I have 5 hearts	I have 10 pears	I have 2 cars	I have 8 stars
Where are 4 hearts?	Where are 3 pears?	Where are 6 cars?	Where are 7 stars?

Count pictures and underline the answers

(apples)	8 6	Eight Six	(ducks)	5 2	Five Two
(cups)	8 6	Eight Six	(pineapples)	2 3	Two Three
(pears)	5 4	Five Four	(leaves)	5 10	Five Ten
(butterflies)	9 6	Nine Six	(flowers)	7 6	Seven Six

Colour the pictures which are in the ordinal numbers place
(from left to right)

First										
Second										
Third										
Fourth										
Fifth										
Sixth										
Seventh										
Eighth										
Ninth										
Tenth										

Fill in the missing places and find the digit words

8 E i g h t	9 N i n e	10 T e n
E _ g _ t	N i n _	T _ n
E _ _ h _	N _ _ e	T _ _
_ _ _ _ t	N _ _ _	_ _ _

Trace the number and word

8	Eight	Eight	Eight
Eight	Eight	Eight	Eight

Circle the pictures which are in the ordinal numbers place
(from down to top)

| First | Second | Third | Fourth | Fifth | Sixth | Seventh | Eighth | Nineth | Tenth |

Connect the digit word in order 1 - 10

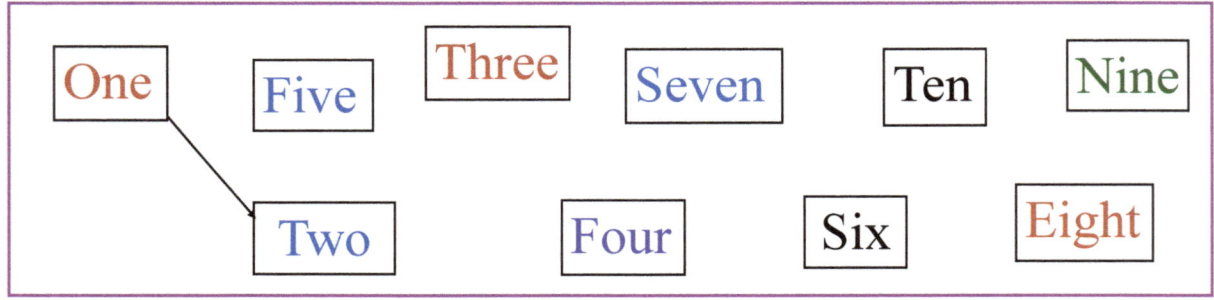

Trace the number and word

9	Nine	Nine	Nine	Nine
Nine	Nine	Nine	Nine	Nine

Count the pictures and fill the numbers and their words in the missing places

	🍎									One
2	🍎	🍎	----------------							
	🍎	🍎	🍎	----------------						Three
4	🍎	🍎	🍎	🍎	------------					
	🍎	🍎	🍎	🍎	🍎	----------				Five
6	🍎	🍎	🍎	🍎	🍎	🍎	--------			
	🍎	🍎	🍎	🍎	🍎	🍎	🍎	------		Seven
	🍎	🍎	🍎	🍎	🍎	🍎	🍎	🍎	----	
9	🍎	🍎	🍎	🍎	🍎	🍎	🍎	🍎	🍎	---
	🍎	🍎	🍎	🍎	🍎	🍎	🍎	🍎	🍎	🍎	Ten

Trace the number and word

10	Ten	Ten	Ten	Ten	Ten	Ten

Colour the squares which are in the place of the ordinal numbers

First										
Second										
Third										
Fourth										
Fifth										
	1	2	3	4	5	6	7	8	9	10

										Sixth
										Seventh
										Eighth
										Nineth
										Tenth
10	9	8	7	6	5	4	3	2	1	